[W]orkers in artificial intelligence —
blinded by their early success [. . .] —
will settle for nothing short of the
moon. [. . .] To persist in such optimism
in the face of recent developments
borders on self-delusion.
Hubert Dreyfus (1965)

No one in 2015 would dream of buying a
machine without common sense, any more
than anyone today would buy a personal
computer that couldn't run spreadsheets,
word processing programs, communications
software, and so on.
Doug Lenat (1990)

The development of full artificial
intelligence could spell the end
of the human race.
Stephen Hawking (2014)

Series 117

This is a Ladybird Expert book, one of a series of titles for an adult readership. Written by some of the leading lights and outstanding communicators in their fields and published by one of the most trusted and well-loved names in books, the Ladybird Expert series provides clear, accessible and authoritative introductions, informed by expert opinion, to key subjects drawn from science, history and culture.

MICHAEL JOSEPH

UK | USA | Canada | Ireland | Australia

India | New Zealand | South Africa

Michael Joseph is part of the Penguin Random House group of companies whose addresses can be found at global.penguinrandomhouse.com

First published 2018

002

Text copyright © Michael Wooldridge, 2018

All images copyright © Ladybird Books Ltd, 2018

The moral right of the author has been asserted

Printed in Italy by L.E.G.O. S.p.A.

A CIP catalogue record for this book is available from the British Library

ISBN: 978–0–718–18875–7

www.greenpenguin.co.uk

Artificial Intelligence

Michael Wooldridge

with illustrations by
Stephen Player

Ladybird Books Ltd, London

The idea of AI

We all know something about Artificial Intelligence (AI). From the deadly computer HAL-9000 in *2001: A Space Odyssey* to the accommodating robot hosts of *Westworld*, countless movies, novels and computer games have enthralled us – and sometimes terrified us – with the prospect of conscious, self-aware, intelligent machines.

But AI is not just science fiction. Since the first computers were developed in the 1950s, AI has been an active scientific discipline, and many developments in modern computing in fact trace their origins to AI research. There have been genuine breakthroughs recently that would have astonished the early AI pioneers. But AI has a notorious track record for overselling itself. All too often, AI researchers have let their excitement and optimism lead to wildly unrealistic predictions of what they were going to achieve. The history of AI is full of ideas that initially showed promise, but which, in the end, didn't work as hoped. Because of this, many people are sceptical about AI.

The reality of AI today – what has been achieved, and what might be possible – is tremendously exciting, but it is far removed from the AI of science fiction. In this book, we will explore what AI really is. Starting from its origins, we will examine the various ideas that have shaped the field, taking us to the present day, when AI systems are all around us. We will then look at where AI might ultimately take us.

The Turing test

One of the first scientists to think seriously about the possibility of AI was the brilliant British mathematician Alan Turing. To all intents and purposes, Turing invented computers in the 1930s, and soon after became fascinated by the idea that computers might one day be intelligent. In 1950, he published a scientific paper on the question of whether a machine could 'think'. The paper introduced the 'Turing test':

You are interacting via a computer keyboard and screen with something that is either another person or a computer program. The interaction is in the form of text – questions and answers. Your task is to determine whether the thing being interrogated is in fact a person. Now suppose, after some time, you cannot tell whether the thing is a person or program. Then, Turing argued, you should accept that the thing being interrogated has human-like intelligence.

Turing's genius was to see that the test sidesteps issues such as *how* the program is doing what it is doing: anything that passes the test is doing something *indistinguishable* from human behaviour.

Ingenious as it is, Turing's test has limitations. For one thing, it looks only at one narrow aspect of intelligence. Also, it is possible to write programs that use cheap tricks to confuse the interrogator (this is what Internet 'chatbots' do – these are not AI). Contemporary AI researchers have developed refined versions of Turing's test, which are resistant to such trickery.

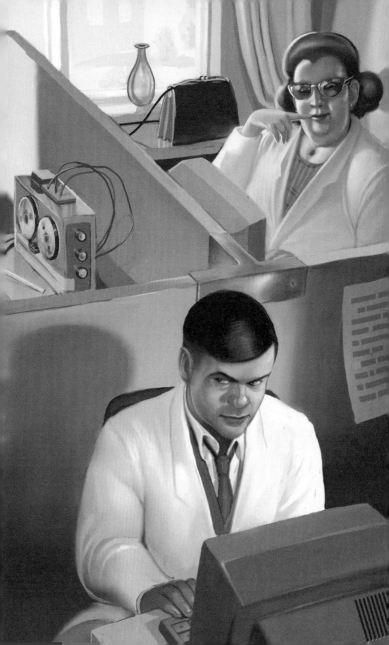

The Chinese Room

Many people have argued that machines can never *really* be conscious or self-aware. One argument is that there is something inherently 'special' about people. It is hard to rebut this claim, but also hard to accept it: ultimately, people are just atoms bumping into each other. A more sophisticated argument against the possibility of AI was proposed by the philosopher John Searle. His 'Chinese Room' scenario supposedly demonstrates that machines cannot have 'understanding':

Imagine a room in which a man, who understands no Chinese, receives, through a slot in the door, questions written in Chinese. When he receives a question, the man carefully follows detailed instructions written in English to generate a response to the question, which he passes back out through the slot. Now suppose the questions and responses are part of a Chinese Turing test, and the test is passed.

Here, the man plays the role of a computer – the instructions he follows to generate a response are the program. Searle argued that there is no understanding of Chinese anywhere here: the man doesn't understand, and the instructions surely don't. So, there is no understanding.

There are many counter arguments to this. The most common is to say that while no *component* of the Chinese Room understands, *the system as a whole* does. After all, if we look at the components of a human brain, we see no understanding; but the brain as a whole surely does understand. The debate rages.

The components of intelligence

The goal of building machines that can demonstrate general-purpose intelligent behaviour (as in the Turing test) is called 'General AI'. It is hard to tackle General AI directly, partly because we have a poor understanding of general intelligence in humans. Early AI researchers therefore focused on building programs that demonstrated some *components* of intelligence, such as the following.

Perception: understanding our environment. We perceive our world through various mechanisms, including the five senses: sight, sound, touch, smell and taste. When we build robots, we can give them sensors that provide analogues of these, but we can also give them senses that people don't have, such as radar. Perception involves interpreting the raw information provided by sensors, and is probably the single biggest challenge in robotics.

Machine learning: learning from and making predictions about data. Developing machine learning usually involves training a program with many examples. Thus, a program to recognize faces might be trained with pictures labelled with the name of the person in the picture.

Problem solving and *planning*: figuring out how to achieve goals using a given repertoire of actions. Playing a board game would be an example: the goal is to win the game; the actions are the possible moves.

Reasoning: deducing new conclusions from existing facts in a robust way.

Natural language understanding: enabling computers to interpret human languages like English and Chinese.

The Golden Age

In 1956, a young American academic called John McCarthy organized a summer school at Dartmouth College in New Hampshire. His aim was to bring together researchers interested in getting computers to do things that seemed to require brains. To describe the school, he chose the term 'artificial intelligence', and the name stuck. This was the beginning of the 'Golden Age' of AI, which lasted until the mid 1970s.

Those present at McCarthy's summer school were hugely influential in the early development of AI. Chief among them were Marvin Minsky, who co-founded the AI Lab at the Massachusetts Institute of Technology (MIT) in Boston; Alan Newell and Herb Simon, who had just written what was arguably the first AI program, called Logic Theorist, and went on to found the AI Lab at Carnegie Mellon University in Pittsburgh; and McCarthy himself, who founded the AI Lab at Stanford University in the heart of what is now Silicon Valley.

These individuals and their laboratories dominated AI in the Golden Age. Computing was in its infancy, so they and their students invented not just the first AI programs but the tools with which to build these programs. (In the mid 1950s, McCarthy invented a programming language called LISP which, incredibly, is still routinely used across the world today.) Throughout this period, they decisively demonstrated that computer programs were capable of all the key components of intelligence. There was enormous optimism about the likely speed of progress, and many grand predictions.

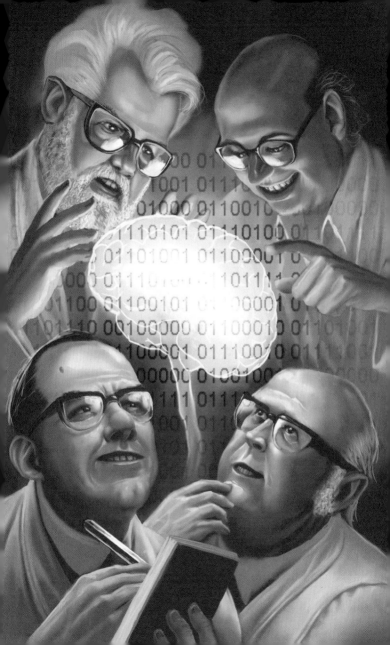

Searching for solutions

'Search' is one of the most fundamental AI problem-solving techniques, and was intensely studied in the Golden Age. In a search problem, we must find a sequence of actions that will take us from some initial state of the world to a goal state. Starting from the initial state, we consider the effects of every available action on that initial state. The effect of performing an action is to transform the world into a new state. If some action generates the goal state, then we have succeeded. Otherwise, we repeat this process for every state we just generated, considering the effect of each action on those states, and so on. In this way, we generate a 'search tree'.

The main difficulty with search is called 'combinatorial explosion'. Put simply, search trees quickly become impossibly large. For example, in a game of chess, there are on average about thirty-five possible moves available from any board configuration, so the final level in a chess search tree just ten moves deep would contain about 3 million billion board positions. A standard technique is to use rules of thumb called 'heuristics' to guide the search process. Heuristics indicate which states are promising, and which are likely to be dead ends.

The crowning achievement for heuristic search came in 1996, when an IBM chess computer called Deep Blue beat reigning world champion chess player Garry Kasparov. Deep Blue processed 200 million positions per second, typically generating a search tree looking ahead six to eight moves.

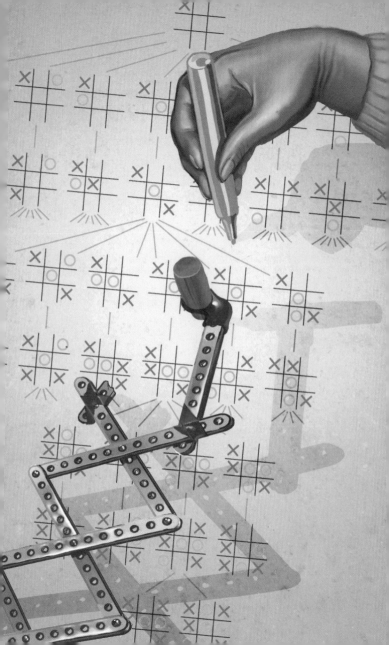

SHRDLU and the Blocks World

The curiously named SHRDLU system was one of the most celebrated achievements of the Golden Age. Developed by Terry Winograd for his 1971 PhD, SHRDLU was based on a scenario called the 'Blocks World'.

The Blocks World was a simulated environment containing a number of coloured objects (blocks, boxes and pyramids). SHRDLU's task was to arrange the objects according to instructions from a user, using a simulated robot arm to manipulate them.

A key feature of SHRDLU was that the user interacted with the system in what seemed to be normal English. Getting computers to interact in natural language had always been a goal of AI, and SHRDLU's dialogue seemed astonishingly rich. However, it later became clear that SHRDLU was able to generate rich dialogues only because those dialogues related to a very constrained scenario – the Blocks World. SHRDLU didn't provide a route to general natural language understanding.

The Blocks World became very popular, partly because it resembles the kind of scenario we might envisage for robots in the real world. However, because it was ultimately a simulated microworld, robotics researchers later argued that it abstracted away most of the really hard problems that a real-world robot would face – in particular, perception.

It is easy for us, fifty years later, to identify the limitations of SHRDLU. But it was an enormous achievement at the time, and hugely influential.

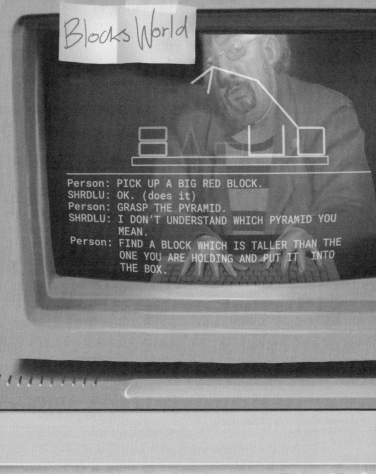

Blocks World

Person: PICK UP A BIG RED BLOCK.
SHRDLU: OK. (does it)
Person: GRASP THE PYRAMID.
SHRDLU: I DON'T UNDERSTAND WHICH PYRAMID YOU
        MEAN.
Person: FIND A BLOCK WHICH IS TALLER THAN THE
        ONE YOU ARE HOLDING AND PUT IT INTO
        THE BOX.

SHAKEY the robot

Another landmark system from the Golden Age was SHAKEY the robot. Developed at Stanford Research Institute, SHAKEY was a serious attempt to build a mobile robot that could be given real-world tasks and would figure out on its own how to accomplish them. SHAKEY was controlled by a computer via a radio link, because the computers of the time were too cumbersome for the robot to carry.

SHAKEY's task was to move objects around an office-like environment. Since SHAKEY had no 'arms', it did this by pushing them. It could perceive its environment using a video camera, rangefinders and bumpers to detect obstacles. SHAKEY's ability to interpret its video feed was very limited – the environment had to be specially painted and carefully lit to provide SHAKEY with a clear view. To control SHAKEY, its designers invented an influential program called STRIPS, which allowed the computer controlling the robot to develop plans to achieve the tasks it was given.

SHAKEY was arguably the first real autonomous mobile robot. But to make SHAKEY work, its designers had to greatly simplify the challenges faced by the robot, and it was far too slow to be of practical use. Although SHAKEY *looked* like the kind of robot that AI had long dreamed of, it demonstrated just how far AI actually was from this goal.

The complexity barrier

Early AI systems demonstrating components of intelligence gave hope that progress would quickly be made on bigger problems. But this hope was not realized. Techniques that seemed promising with microworld scenarios like the Blocks World did not scale up to real-world problems. A mathematical theory of computer problem solving called 'computational complexity' explained why this was the case.

In the early 1970s, Stephen Cook, Leonid Levin and Richard Karp identified a class of computational problems that are now called 'NP-complete' (for 'Non-deterministic Polynomial time complete'). These are problems for which computers can quickly check whether a solution is correct, but for which *finding* a correct solution seems to require an impossibly large amount of time. The 'Travelling Salesman Problem' is a famous example:

A salesman must visit some cities in his car, eventually returning to his origin. His car only has a certain amount of fuel. Is there a route that will complete the tour without running out of fuel?

The best we seem able to do with NP-complete problems is to exhaustively consider all possible solutions. If our salesman had to visit seventy cities, he would have to consider more routes than there are atoms in our universe. No matter how fast computers become, such an exhaustive approach will never be feasible for NP-complete problems.

Unfortunately for AI researchers, seemingly every problem of interest to them proved to be NP-complete (or worse). The Golden Age ground to a halt.

Knowledge is power?

Disappointment with the failure of AI to progress beyond microworld scenarios led to a backlash in the mid 1970s, with research funding cut and widespread recriminations. This miserable period became known as the 'AI Winter'. But by the late 1970s, a new idea was sweeping across the field, which it was hoped would overcome the problems that had dogged AI through the Golden Age.

The new idea was simple enough: explicit knowledge of a problem is the key to tackling complexity. For example, a chess grandmaster does not start from scratch every time she plays a game. Rather, she uses extensive knowledge of the game – what strategies work in which circumstances, and so on. This knowledge helps her to focus her problem solving, avoiding dead ends and directing her to good solutions quickly. To make AI systems work, it was argued, it was necessary to explicitly capture and use this kind of knowledge.

A new class of AI systems began to emerge. These 'expert systems' made use of human expert knowledge to solve problems in tightly constrained areas. Expert systems were not General AI, but they seemed to provide evidence that AI systems could outperform humans in solving some problems, and for the next decade, knowledge-based systems remained the main focus of AI research.

MYCIN

MYCIN was one of the most famous systems from the days of knowledge-based AI, and a classic example of an expert system. MYCIN was intended to be a doctor's assistant, providing expert advice about blood diseases.

MYCIN's knowledge about blood diseases was represented in the form of about 200 rules. Each rule captured a discrete piece of knowledge about blood diseases, and was carefully written after discussions with human experts. This task – extracting knowledge from human experts – is difficult and time-consuming, because people often can't articulate the knowledge they have. A typical MYCIN rule (expressed in English) is as follows:

IF: 1) The organism does not stain using the Gram method **AND**
2) The morphology of the organism is rod **AND**
3) The organism is anaerobic

THEN:
There is evidence (0.6) that the organism is bacteroides

MYCIN had many features that came to be regarded as essential for expert systems. First was the need to be able to explain a conclusion, so that human users could understand it. Second was the need to cope with uncertainty – most conclusions, as in the rule above, were not crisp yes/no answers.

Trials indicated that MYCIN performed better in its area of expertise than human specialists – an impressive and much-lauded feat. But its expertise was very narrow – it could operate *only* in its area of expertise.

My preferred therapy recommendation is as follows: Give the

following in combination:

ETHAMBUTAL

Dose: 1.289 (13.0 100mg-tablets) q24h PO for 60 days

[calculated on basis of 25 mg/kg] then 770 mg (7.5 100mg-

tablets) q24h PO [calculated on basis of 15 mg/kg]

The dose should be modified in renal failure. One can adjust

the dose only if a creatinine clearance or a serum creatinine

is obtained.]

Comments: Periodic vision screening tests are recommended for

optic neuritis.

CYC: The ultimate expert system?

CYC was the boldest experiment from the heyday of expert systems. Its inventor, Doug Lenat, became convinced that knowledge provided the route to General AI, but that we needed more of it. Much, *much* more. He proposed to build CYC: a system with all the knowledge that a reasonably educated person uses as they go about their life. And there was no quick way to get this knowledge, Lenat argued: it all had to be laboriously encoded, just as MYCIN's knowledge about blood diseases was encoded.

This was mind-bogglingly ambitious. Think about all the knowledge you have about the world. CYC would need to know that you cannot eat Kansas; that trees are plants; that when you drop something, it falls; that eating too much is usually bad for you; that China is a country; that cats are usually pets, but not always; that red-labelled taps usually provide hot water; and so on.

Lenat was famously optimistic about CYC. He believed that after a while CYC would know enough to be able to educate itself. More than thirty years later, Lenat still works on CYC, and it has been spun off into a commercial company. But CYC never demonstrated anything like the capabilities that Lenat expected and hoped for, and for this reason CYC is often cited as an example of the AI community's unrealistic optimism. But perhaps CYC hasn't failed. Perhaps it just needs a few more rules . . .

Logical AI

As the ambition of knowledge-based AI researchers grew, they looked for richer and more precise ways to capture knowledge. Mathematical logic was an obvious candidate.

Mathematical logic was developed to understand reasoning, and is very effective for this. In AI, the idea was that logic would lend mathematical rigour to the process of representing and using knowledge. Intelligent decision-making, it was hoped, could then be reduced to purely logical reasoning. The spirit of logical AI was captured in a popular programming language called PROLOG. In PROLOG, a programmer expresses their goal and their knowledge about a problem in a simple logical form, and the computer tries to solve the problem using logical deduction.

But many difficulties with the logical approach to AI soon became apparent. One notorious problem was 'common-sense reasoning':

You are told Tweetie is a bird. You conclude that Tweetie can fly. Later, you are told that Tweetie is a penguin, so you retract your conclusion.

The type of logic used in mathematics can't cope with this seemingly trivial scenario, because it wasn't designed for retracting conclusions. Moreover, logical reasoning turned out to be very hard to automate. And more generally, logical reasoning is simply inappropriate for many tasks – driving a car, for example, doesn't seem to require much logical deduction.

Logic and reasoning proved to be powerful tools for some problems, but did not provide a route to General AI.

Coping with uncertainty

The Turing test established human behaviour as the goal of AI. But humans are often poor decision-makers – why should we want to build systems that make bad decisions? The goal of AI began to shift from making *human* decisions to making *rational* decisions.

A key aspect of rational decision-making relates to reasoning with uncertain information. People are bad at this. Consider:

A deadly new flu virus infects one in every thousand people. A test for the flu is developed, which is 99 per cent accurate. On a whim you take the test, and it comes out positive.

Most people would be very worried, but paradoxically there is only about a one in ten chance that you have the flu. Why? Because the prior probability that you had the flu was one in a thousand – far more people *don't* have flu than *do*, so there will be many more false positives than true positives.

The basic mathematics behind this reasoning was developed by the Reverend Thomas Bayes in the eighteenth century, but much work was needed to make Bayesian reasoning usable in AI, because AI systems often have to deal with *lots* of structured evidence.

Many automated translation systems use Bayesian reasoning. They compute the likeliest translation of a word, given the words that have appeared previously. Prior probabilities are computed by examining many translated texts. These translation tools are very useful for routine tasks, but in no sense do they *understand* the text they translate.

Out of 1000 people, 11 will test positive for flu, but only one will have it.

Nouvelle AI and the robotics revolution

Knowledge-based and logical AI delivered many successes, but these were mostly in the form of MYCIN-style expert systems, capable of solving problems only in narrow, very well-defined areas.

This led to frustration in the mid 1980s, with some researchers beginning to question the direction AI research had taken from the beginning. One widely heard opinion was that it was a mistake to focus on disembodied systems like MYCIN, which are not based in and acting on the real world. The real proving ground for AI, it was argued, should be the physical world. There was a resurgence of interest in robotics as a consequence.

The most outspoken and influential advocate of robotics-based AI in this period was Rodney Brooks. He became disillusioned with the then-prevailing paradigm of knowledge-based AI, and came to be convinced that intelligent behaviour could be generated without explicit knowledge and reasoning of the kind promoted by knowledge-based AI in general, and logic-based AI in particular.

For many, Brooks's arguments crystallized their increasing discomfort with the long-standing orthodoxy of AI. Of course, some old-time AI researchers were sceptical. Someone coined the terms 'nouvelle AI' to describe the new ideas, and 'good old-fashioned AI' to describe the established tradition. There was a schism in the AI community, which has never healed, between those who emphasize knowledge and reasoning and those who believe these are not the right foundation for AI.

Behavioural AI

The idea of behavioural AI is to focus on the individual behaviours that we expect an intelligent system to exhibit, and then to think about how these behaviours are related. From the late 1980s, Brooks and his colleagues developed a series of autonomous robots, embodying this new paradigm.

Consider a cleaning robot, which must explore some environment picking up rubbish. Such a robot, Brooks argued, doesn't need sophisticated knowledge and reasoning to do its job effectively. Instead, an efficient robot can be made by combining a few extremely simple behaviours:

- If you detect an obstacle, change direction.
- If you are carrying rubbish and are by a bin, then dump the rubbish.
- If you detect rubbish, pick it up.
- Move randomly.

These behaviours are arranged in the robot so that the first has the highest precedence, and the last has the lowest (so if it ever detects an obstacle it will always change direction, while it will move randomly only if none of the other behaviours is active). The approach was influential, although it later became clear that it was difficult to apply the approach to more than a few behaviours.

You may have encountered a robot whose design was influenced by these ideas. In fact, you may well have one in your home: Brooks was a founder of the company iRobot, manufacturer of the popular Roomba robot vacuum cleaners, the design of which was based on his work.

Move randomly

if carrying rubbish and by bin **then** dump rubbish

if detect rubbish **then** pick it up

if detect obstacle **then** change direction

sensors

actuators

Autonomous vehicles: the DARPA Grand Challenge and robot STANLEY

The potential benefits of computer-controlled vehicles made them an important subject of research far beyond the field of AI. Globally, more than a million people die in automobile accidents every year: computer control has the potential to save lives on a massive scale.

By 2004, progress was such that the US military defence research agency DARPA organized a 'Grand Challenge', inviting researchers to enter a competition in which vehicles would autonomously traverse 150 miles of the US countryside. The results highlighted just how hard the problem was: none of the fifteen finalists completed more than eight miles of the course.

Undaunted, DARPA organized a second Grand Challenge. The 195 entries were whittled down to 23 finalists, who competed on 8 October 2005, attempting to cross 132 miles of Nevada desert. This time, five teams completed the course. The winner was the robot STANLEY, designed by a team from Stanford University led by Sebastian Thrun. STANLEY completed the course in just under seven hours, averaging about 20 mph. A converted Volkswagen Touareg, STANLEY was equipped with seven onboard computers, interpreting sensor data from GPS, laser rangefinders, radar and video feed.

The age of driverless cars dawned on that day. The achievement was every bit as important as that of the Wright brothers at Kitty Hawk a century earlier – and driverless vehicles will change our world every bit as much as did their invention of the aeroplane.

The modern era, and the rise of machine learning

The modern era of AI began around 2005, and has been characterized by dramatic and highly publicized advances in one AI technology in particular: machine learning.

Machine learning aims to build computers that can learn how to do things without being explicitly told how. Machine-learning systems require training in order to learn. In 'supervised learning', training takes place by presenting a program with examples of the thing that the computer is trying to learn. Face recognition software is the most obvious example: when you identify individuals in a photo on social media, you are providing training data for machine-learning algorithms, so that those algorithms will be able to identify those individuals.

In 'reinforcement learning', a system is able to experiment by making decisions, and receives feedback on those decisions (whether they were good or bad). If a system receives feedback that a decision was bad, it will be less likely to make that decision in the same circumstances in future.

There has been impressive progress with machine learning recently, prompted by three developments. First, there were scientific breakthroughs with 'deep learning' systems that can cope with complex problems. Second, to make machine learning work reliably, it turned out that lots and lots of data was necessary for training – and data, from endless sources, is now abundant. And finally, training requires lots of computer processing power, which is now available very cheaply.

A machine-learning program is trained by giving it examples of the thing it is supposed to recognize.

Neural nets

A popular approach to machine learning uses 'neural nets', in which many small artificial neurons are connected in complex networks. Each neuron takes inputs from its neighbours, and generates an output depending on certain numeric weights associated with these inputs. The output will then in turn influence the neuron's neighbours. By adjusting the weights, the system can be made to learn associations between inputs and outputs. Although neural nets were inspired by the microstructure of the brain, it is important to understand that neural net researchers are not trying to build artificial brains.

Although neural nets had been studied since the beginning of AI, early research on this approach came to an abrupt end in the 1960s, when Marvin Minsky and Seymour Papert proved that there are severe limits to what simple neural networks can do. The field was dormant until a revival began in the 1980s, prompted by the observation that richer models, called 'parallel distributed processing', could overcome these problems. There has been rapid progress this century.

A key problem with neural nets (and machine learning generally) is that the intelligence they embody is usually opaque. For example, a neural net that has been trained to recognize cancerous growths on X-ray pictures cannot explain its decisions. The expertise the system has is hidden in the numeric weights associated with neurons, and there is no easy way to extract the knowledge that these weights implicitly carry.

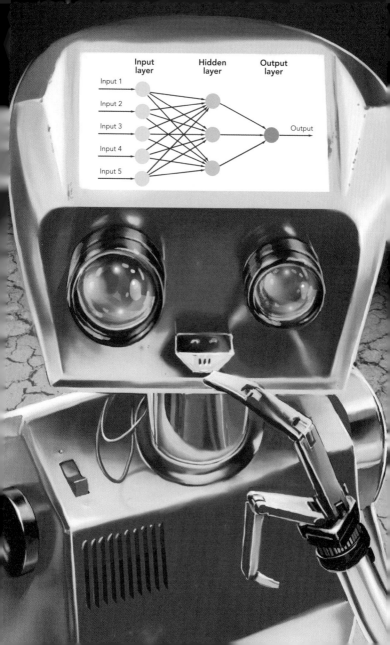

DeepMind and Alpha Go

Progress in machine learning is perhaps most vividly epitomized by the story of the UK AI company DeepMind. Formed in 2012, DeepMind were acquired by Google in January 2014 for the reported sum of $400 million. At the time, the company were virtually unknown, even within the AI community – they apparently had no customers and only a handful of employees.

Later in 2014, DeepMind showed the world why Google were so interested in this tiny company. They demonstrated a system that had learned to play forty-nine arcade games from a 1980s Atari video console, ultimately playing twenty-nine of these at above-human levels of performance. The inputs to the system were simply the video feed (i.e., what a person would see) and the current score, and the controls were the same as humans use. Crucially, the program was given no knowledge of the game at all. It learned to play the games using reinforcement learning combined with neural networks. The system made choices, got feedback and adjusted its choices accordingly. One single program learned to play all the games, just through this process of trial and error.

This was a striking achievement – but there was more to come. In March 2016, a DeepMind program called Alpha Go decisively beat a human champion in the board game Go over a series of five matches in Seoul, Korea. Go is *much* harder to play than chess, and no serious progress in computer Go had been expected for decades. The achievement made front-page news across the world.

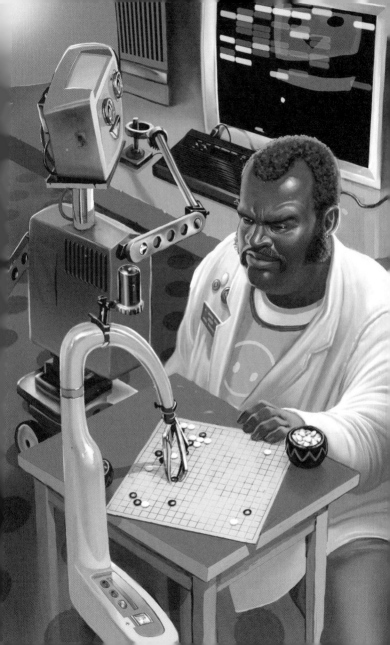

AI today – AI everywhere

AI is not something that awaits us in the future. You encounter it every day. Digital assistants like Siri, Alexa and Cortana are classic AI. The software that recognizes faces in the photos you upload to social media is AI. The software in your car's satellite navigation system is AI. The product recommendations you receive from online stores are made by AI. Cruise control systems in modern cars are AI. When you apply for a loan from a bank, AI may be involved in the decision about whether you receive it; the same bank might use AI-based automated trading systems. AI-based automated translation systems are routinely used across the world.

AI today is everywhere, and it will be ever more prominent in the years to come, because AI software can reliably and efficiently make better decisions than people in a huge range of settings.

In the future, AI will be invisibly embedded everywhere that decisions are made. One way to think about this is as follows. In the early part of this century, satellite navigation systems revolutionized driving, by automating the tedious and error-prone task of navigation. Satellite navigation systems act like a prosthetic device for the brain: they take the strain of a complex cognitive task. Now imagine such cognitive prosthetics assisting you with every decision you have to make.

This is a long way from General AI, but it is the likely reality of AI for the foreseeable future.

Jane Doe
Flight 2206
London-Leeds
arriving on time
Rent
8999
Green.
Restaurant
Rating
3 out of 5

Concerns

Ubiquitous AI will surely bring many benefits, but it will also present challenges for society and law-makers.

The most obvious of these is unemployment. For example, there are 3.5 million truck drivers in the USA: most of these jobs will become automated in the decades ahead. Although this seems alarming, it should be understood as part of a long-term trend towards the automation of human labour that started with the Industrial Revolution in the eighteenth century. While automation initially took the jobs of unskilled labourers, AI will take ever-more skilled roles.

Privacy is another concern. For example, computers are better at recognizing faces than people, and can accurately predict our sexual orientation and political views from our social media feeds.

Another concern is *algorithmic bias*. While AI decision-makers can in principle be designed to be free of the (conscious or unconscious) biases that all of us have, they are only as good as the humans who design and train them. Poorly designed algorithms will make poor decisions. And if a machine-learning program is trained by someone with biases, then that program will also be biased.

Finally, many people are profoundly uncomfortable with the idea of autonomous weapons, which have the power to decide whether to take human life. While it has been argued that autonomous weapons can be designed to be more reliably ethical than humans, many people find the very idea of autonomous weapons to be abhorrent.

The singularity and all that

A long-term concern is what happens if we reach the 'singularity' – the hypothesized point at which General AI systems become smarter than people. The worry is that these systems might use their intelligence to improve themselves, then apply their smarter selves to improving themselves further, and so on. Perhaps AI will then be beyond human control, and may even pose a threat to our existence. Scenarios like this are a staple of science fiction, familiar from movies such as *Terminator* and TV shows like *Battlestar Galactica*.

In discussions about the singularity, someone usually suggests that we need the 'Three Laws of Robotics', proposed by celebrated science fiction author Isaac Asimov:

1 A robot may not injure a human or, through inaction, allow a human to come to harm.
2 A robot must obey orders given by humans except where they would conflict with the First Law.
3 A robot must protect its own existence as long as this does not conflict with the First or Second Laws.

Unfortunately, Asimov's Laws require a robot to be able to predict the consequences of all its actions into the distant future. This is too computationally challenging to ever be possible.

Scientific opinion is divided about whether the singularity might happen. Even if it did, it is not obvious why the nightmare scenario should occur. There is much uncertainty about the long-term future of AI. For the foreseeable future, though, it is a safe bet that far more people will die from natural stupidity than from artificial intelligence.

Conscious machines

Recent advances in AI, impressive as they are, are narrow achievements. They provide no direct route to General AI or the dream of conscious intelligent machines. In part, this is because human consciousness and intelligence remain poorly understood.

We are – slowly – beginning to understand consciousness. The field of cognitive neuropsychology has begun to shed light on how the physical brain gives rise to psychological processes. Devices such as Magnetic Resonance Imaging (MRI) systems allow us, for the first time, to look inside brains and see what is happening as they do their work. Evolutionary psychology gives us clues about when and why certain features of human consciousness appeared, and therefore the role that consciousness plays.

Although there is no obvious route to General AI, there is no evidence that it is impossible. There is nothing physically special about people and their brains – we don't yet understand them, but they are not magical. Conscious, self-aware machines are surely possible, at least in principle. But *our* consciousness is the product both of millions of years of evolution and of our experiences as individuals in the human world. Our brains are monkey brains, inhabiting monkey bodies, and human consciousness is surely determined by this. Machine consciousness will, therefore, be unlike our own.

We are still a long way from understanding consciousness in humans, but ultimately we will understand it, and when we do, machine consciousness will not seem such a startling and unlikely prospect.

Glossary

BAYESIAN REASONING: A way to adjust your beliefs about the world when given new data or evidence; the main method in AI for reasoning with uncertain information.

BLOCKS WORLD: A simulated 'microworld' in which the task is to arrange various objects like blocks and boxes.

EXPERT SYSTEM: A system that uses human expert knowledge to solve problems in a tightly constrained area.

GENERAL AI: The grand dream of AI: self-aware, conscious, intelligent machines.

GOLDEN AGE: The early period of AI research, from about 1956 to 1975 (followed by the 'AI Winter').

GRAND CHALLENGE: A competition for driverless cars, which led to the triumph of robot STANLEY in October 2005 and the dawn of the age of driverless cars.

KNOWLEDGE-BASED AI: The dominant paradigm for AI from about 1975 to 1985, which focused on using explicit knowledge about problems, often in the form of rules.

MACHINE LEARNING: Programs that learn how to do things.

NATURAL LANGUAGE UNDERSTANDING: Programs that can interact in ordinary human languages like English.

NEURAL NETS: An approach to machine learning using 'artificial neurons'.

NOUVELLE AI: Starting about 1985, the rejection of knowledge and reasoning, in favour of a focus on behavioural approaches.

NP-COMPLETE: A class of computational problems that resist attempts to solve them efficiently. Many AI problems are NP-complete.

PERCEPTION: Understanding what is around you in your environment.